The Creation and Sustaining
of the
Universe and All Life on Earth

THE CREATION AND SUSTAINING OF THE UNIVERSE AND ALL LIFE ON EARTH

A CURRENT EVIDENCE-BASED ESSAY

With Essay Background:

The Evolution of Anti-Evolution in Public-School Science Classes

Glen E. Robinson, Ph.D.

Published by G.E. Robinson & Company.
U.S. Post Office, P.O. Box 1866
Davidson, NC 28036

Printed in United States of America

Library of Congress Catalog Number 2013902712

Standard ISBN number 978-0-578-11861-1

Design and Layout by SelfPublishing.com

This publication is available from:

Amazon.com, BarnesandNoble.com, IngramBook.com

and other book sources.

Table of Contents

Background of Essay:
*The Evolution of Anti-Evolution in
Public-School Science Classes*

FOREWORD

Glen E. Robinson is President Emeritus of Educational Research Service, the research and information arm for school management in the US and Canada. He is a WWII combat infantry officer veteran who served with the 10th Mountain Division in Italy. After the war, he completed B.S. and M.S. degrees at University of Tennessee and then in 1949 became the Science teacher in Norris School, a 12 grade public school in Norris, Tennessee.

Norris is a small town about 70 miles from Dayton, Tenn. where only 25 years earlier, Science teacher John Scopes was tried, found guilty and fined for teaching the theory of Evolution in violation of the state anti-Evolution law. In mid-July 1925, the famous Scopes Trial, with nationally known lawyers and a host of news photographers, was the first public trial ever to be broadcast on public radio. The trial was sensationally publicized both here and abroad in newspapers of the day as the "Scopes Monkey Trial".

As the new and only Science teacher in this highly charged anti-scientific environment, Robinson was repeatedly confronted by student questions about Evolution. The challenge was to respond to student questions in ways that were honest and scientifically accurate yet were not overt violations of the Tennessee anti-Evolution law in ways that could be used as evidence against the new Science teacher. Some persons and groups in the locality were known to be poised for a local replay of the Scopes Trial episode.

But his basic challenge as a Science teacher was to help students see beyond the controversy over how the Earth and life originated and to help them become curious, excited learners seeking to understand and appreciate the Cosmos and the many interesting forms of life on Earth, and to appropriately use Scientific methods of thinking and reasoning to sort-out truth from untruth.

During his second year at Norris, he was recalled to active duty as an infantry officer during the Korean War. He later returned to post-graduate study and received the PhD from George Peabody College, now part of Vanderbilt University, in Nashville. He became a faculty member and Assistant to the President of Peabody College.

During the turbulent period in the South following the U.S. Supreme Court's 1954 decision to racially desegregate southern schools "With all deliberate speed", he wrote many published accounts of the important and sometimes heroic roles that public school superintendents and principals were playing in implementing the Supreme Court and lower court decisions. One of the most notable of those accounts was the final section of the 1957 book titled *With All Deliberate Speed*, edited by Don Shoemaker. The book summarized the first three years of desegregation conflict. Robinson called the section "The Man in No Man's Land." Southern school administrators were literally the person caught in the middle when they tried to implement a court order to desegregate public schools in a community that was avidly determined to preserve racial segregation.

Later, the Robinson family moved to the Washington, DC area where he became Director of the National Education Association's (NEA) Research Division. There he pioneered the development and use of scientific sampling and computer processing of data and information in many areas of education. With the advent of collective bargaining in public employment, the school administrator associations withdrew as NEA affiliates and asked Robinson to establish and direct a new, independent, nonprofit Educational Research Service (ERS) with the mission to provide timely and reliable research and information needed for effective management of the nation's public schools. He established the ERS in 1973 and served as its President and Director of Research until his retirement as ERS President Emeritus in 1995.

While living in Bethesda,MD for 47 years,he and his family (wife Mary, 3 sons and daughter) were active members of the North Bethesda United Methodist Church. For several years, the author taught the church's Sunday School Class for high-school youth.

Many of the concepts and ideas in this Essay are the result of the author's research and concern for helping those alert and thoughtful youth to understand and appreciate that the belief God created the Universe and life on Earth does not require belief that all creation was accomplished in six days by divine command.

Rather,belief that God created the heavens and life on Earth calls for sincere and diligent efforts to learn and appropriately consider scientific evidence of the amazing ways and means used to create,sustain and change the Cosmos and life on planet Earth.

The author sincerely hopes that the concepts and information described in this Essay will be helpful to the youth of today and to others.

*

Glen and Mary Falls Robinson were married in 1949 and now live at *The Pines at Davidson*, an excellent retirement community,in Davidson,NC.–home of Davidson College.

WHY THIS ESSAY

Controversy has long raged over one of the most profound questions of our time. *Which of two ways was the Universe and all life on Earth created and continues to operate?* One is called **Creationism** and the other called **Evolution**.

Creationism is the theology-based belief that the Universe and all Life on Earth was created by God in the **brief time** of literally one, six-day work-week by use of Divine commands and then a seventh day of Divine rest.

Evolution is the science-based theory that Creation of the Universe and all life on Earth took place, are being sustained and changed over **extended time** by natural laws and principles of nature. (Also called Darwinism.)

This controversy has continued for 150 years and much has happened during that time. Yet, most arguments in this current DNA-Space-Age Era are much the same as those used a century ago.

There is need for a fresh and current examination of the information now available that pertains to these issues.

This Essay is a sincere effort to present fairly, justly and objectively past and current evidence pertaining to **both ways** that could have been used in Creation, so that readers who are truly interested in learning **how** the Cosmos and all life on planet Earth were created and continue to operate can decide for themselves how Creation was actually accomplished.

To help understand the important need for a fresh and current examination of the information now available pertaining to these issues, a brief *Background Summary* is appended to this Essay, titled: *The Evolution of Anti-Evolution in Public-School Science Classes.*

In order to prevent the purpose and origin of this Essay from being mistaken, the author states that he is solely responsible for the contents, constructs, and format of the Essay. There have been no ghost or rewriters involved. The author's old notes from long-ago science-teacher and Sunday-school teacher days have been used appropriately. Current source materials have been obtained by the author's library searches and online searches on his home computer. The final manuscript has been independently published and paid for by the author.

Special recognition and much gratitude is due Mary Falls Robinson, wife of the author, for her devoted assistance and editing of this and many other manuscripts during their 64 years of marriage.

The Essay style was specifically used in this writing to facilitate the rapid compilation and descriptive use of evidence from varied sources pertaining to many topics in the shortest time and space possible. Thus, for the sake of brevity and speed of coverage of this large topic, the Essay does not have the precise documentation used in formal research writing and analysis. The author wishes to acknowledge this limitation and, more importantly, to express his deep appreciation to the many persons and sources from which evidence and facts in the Essay have been used, always in the context of their original source.

Finally, the 89 year old author, WWII veteran, former public school Science teacher, former Sunday school teacher of high-school youth, and President Emeritus of Educational Research Service, dedicates this Essay especially to the current generation of American youth, so that they might decide for themselves how the Universe and life on Earth were created, being sustained and changed, and thus be better prepared to live, learn, enjoy and participate understandingly in this fascinating scientific age.

—Glen Robinson

The Creation and Sustaining of the Universe and All Life on Earth

A Current Evidence-Based Essay

To find answers to the fascinating and important question: *How was the Universe and all Life on Earth Created and Sustained?*, let us start with a basic premise and a firm commitment.

- The basic **premise** is that the Cosmos, the Earth, and all the myriad forms of life on Earth do now exist and that in some way they were all created.

- The **commitment** is that we will learn as much as we can about **how** all that is in nature and the Universe was created.

The basic question: *What evidence is available on the creation and sustaining of the Universe and all life on Earth?*

EVIDENCE FROM
THE HOLY BIBLE

For answers to the basic question, let us first examine **all** the evidence contained in *The Holy Bible* (Revised Standard Version, Second Edition, 1971):

1) **Genesis 1.1-2.3, states:**

"In the beginning God created the heavens and the Earth. The earth was without form and void, and darkness was upon the face of the deep; and the spirit of God was moving over the face of the waters. And God said, 'Let there be light'; and there was light; and God saw that the light was good; and God separated the light from the darkness. God called the light Day and the darkness he called Night. And there was evening and there was morning, one day."

On the 2nd day, God separated the waters to above and below a firmament. "And God called the firmament Heaven."....

On the 3rd day, "God said, 'Let the waters under the heavens be gathered together into one place and let the dry land appear.' And it was so. God called the dry land Earth, and the waters that were gathered together he called Seas."

Then on the 3rd day, "God said, 'Let the earth put forth vegetation, plants yielding seed, and fruit trees bearing fruit in which is their seed, each according to its kind, upon the earth.'"....

On the 4th day "God said, 'Let there be lights in the firmament of the heavens to separate the day from the night; and let them be for signs for seasons and for days and years, and let them be lights in the firmament of the heavens to give light upon the earth.' And it was so. And God made the two great lights, the greater light to rule the day and the lesser light to rule the night; he made the stars also." ...."

On the 5th day "God said, 'Let the waters bring forth swarms of living creatures, and let birds fly above the earth across the firmament of the heavens.' So God created the great sea monsters and every living creature that moves, with which the waters swarm, according to their kinds, and every winged bird according to its kind."....

On the 6th day, "God said, 'Let the earth bring forth living creatures according to their kinds: cattle and creeping things and beast of the earth according to their kinds.' And it was so."....

"Then God said, 'Let us make man in our image, after our own likeness; and let them have dominion over the fish of the sea, and over the birds of the air, and over the cattle, and over all the earth, and over every creeping thing that creeps upon the earth.' So God created man in his own image, in the image of God he created him; male and female he created them."

"God blessed them and God said to them, 'Be fruitful and multiply, and fill the earth and subdue it; and have dominion over the fish of the sea and over the birds of the air and over every living thing that moves upon the earth.'"….

"And God saw everything he had made, and behold, it was very good. And there was evening and there was morning, a sixth day."

On the 7th day, creation is declared finished. "Thus the heavens and the earth were finished, and all the host of them. And on the seventh day God finished his work which he had done, and he rested on the seventh day from all the work which he had done. So God blessed the seventh day and hallowed it, because on it God rested from all his work which he had done in creation."….

It should be noted that in this six-day account of how God created the heavens and life on Earth that day and night were created the first day, plants and trees were created the third day, but the Sun and Moon were not created until the fourth day.

In general, this striking account of creation might be viewed as God revealing to humankind as much of God's amazing creative processes as people could see and understand without the aid of telescopes, microscopes, or other technologies.

The six-day description of creation contained in the first chapter of Genesis should be appropriately recognized for the tremendous role that this account of creation has played in providing answers that have satisfied the concerns of millions of people through the ages about how the heavens, the Earth, and all life were created. It also deserves recognition for a number of other things, including our seven day week calendar, the religious Sabbath, and the six day workweek even in societies that sanctioned slavery.

2) **Genesis 2.4—3.25**

Genesis 2 presents a much different account of the creation of life on Earth than the water account in Genesis 1. The Genesis 2 account of creation begins with a dry desert-like environment:

"In the day that the Lord God made the earth and the heavens, when no plant of the field was yet in the earth and no herb of the field had yet sprung up—for the Lord God had not caused it to rain upon the earth, and there was no man to till the ground; but a mist went up from the earth and watered the whole face of the ground—then the Lord God formed man of dust of the ground, and breathed into his nostrils the breath of life; and man became a living being."

The focus of this account is on the first man, Adam, and his wife Eve, and how sin entered the world through their original transgression and description of the punishment that followed.

First, note the name change. In the Genesis 1 account, the creator is "*God*" and all that is created during each of the six days is pronounced "**good**". On the sixth day, God saw everything he had made and pronounced it "**very good**."

In the Genesis 2 account, the creator is "*the Lord God*" and the process of creation proceeds in an opposite, unsatisfactory, trial-and-error kind of way.

The Lord God made the earth and the heavens in one day and first formed man from dust of the ground before forming any plants, animals or birds.

"And the Lord God planted a garden in Eden, in the east; and there he put the man whom he had formed."

*

"Then the Lord God said, 'It is not good that the man should be alone; I will make him a helper fit for him.'"

"So out of the ground the Lord God formed every beast of the field and every bird of the air and brought them to the man to see what he would call them; and whatever the man called every living creature, that was its name.....but for the man there was not found a helper fit for him. So the Lord God caused a deep sleep to fall upon the man, and while he slept took one of his ribs and closed up its place with flesh; and the rib which the Lord God had taken from the man he made into a woman and brought her to the man. Then the man said,

> 'This at last is bone of my bones and flesh of my flesh; she shall be called Woman, because she has been taken out of Man.'"....

3) **The Gospel According to John 1.1-1.5**

John 1 presents still other evidence of how God created the world. It is the prologue to John's gospel establishing the divinity of Jesus and his relationship to God his Father.

"In the beginning was the Word and the Word was with God, and the Word was God. He was in the beginning with God; all things were made through him, and without him was not anything made that was made. In him was life, and the life was the light of men. The light shines in the darkness and the darkness has not overcome it."

These three references are all of the evidences that address directly and factually the creation of the Universe and life on Earth that are contained in *The Old* and *New Testaments* of *The Holy Bible*, Revised Standard Version, Second Edition, 1971. A few passages (Psalms

104, Proverbs 8, and Job 38) refer to the beginning of the earth in poetic or allegorical form; and throughout the Bible, there are many passages pertaining to the power and nature of God, such as, John 4:25 when Jesus said to the Samaritan woman at the well: "God is spirit, and those who worship him must worship in spirit and in truth." But none address directly and factually the creation of the Universe and life on Earth.

*

The concept of **Intelligent Design** requires specific attention. In 1987, the U.S. Supreme Court found that *Creation Science* was not based on scientific evidence but was a religious-based belief. Therefore, it is a violation of the time-honored Constitutional provision of **Separation of Church and State** to teach religious-based **Creationism** in public school science classes.

An alternate approach was constructed by Creationism proponents in the attempt to remove any reference to religious theology from the issue and to term the concept **Intelligent Design**. The approach was that the Universe is so vast, orderly and complex that it had to have been created according to **Intelligent Design** and thus, implying an anonymous **Intelligent Designer**.

In 2005, the U.S. Federal Court in Pennsylvania ruled that *Intelligent Design* was not fact-based theory of the origin of the Universe but a **religious-based belief** similar to that of *Creation Science* and could not be taught as a science-based theory in public school science classes in violation of the **Establishment Clause** of the First Amendment. The court also found specifically that *Intelligent Design* "cannot uncouple itself from its creationist, and thus religious, antecedents."

Promoters of **Intelligent Design** have recently dropped the word *Intelligent* and are now calling the religious-based belief *Design Theory*, and are still actively promoting the teaching of both in public school science classes.

It is important to recognize that many members and groups of both Christian and Jewish faiths no longer believe the literal interpretation that God created the Universe and life on Earth literally in a brief period of time by Divine command. Many now believe that the Universe and life on Earth were created over extended time and by natural means.

*

During the past two centuries, there have been concerted efforts and much progress in the scientific study of natural phenomena. Scientists have been systematically investigating and studying various events, conditions, circumstances, occurrences, and experiences using scientific methods of objective thinking and exploration to learn true cause-and-effect relationships of natural happenings and conditions.

*

Evidence from Science

Now let us examine briefly some of the **scientific evidence** that has been compiled by many persons using their intelligence, determination and **scientific procedures** to learn about **how** the Universe and life on Earth were created, are being sustained, and continuing to change.

*

First we should recognize that scientific **evidence** and **conclusions** are always subject to **skepticism** and **change** in light of subsequent findings and new information.

For example, the current prevailing hypothesis of the origin of the Universe called the "Big Bang Theory"— although long accepted by most astrophysicists—seems to have a basic flaw. In postulating that the origin of the universe began with **one gigantic cosmic explosion** of extremely compact matter and energy, the "Big Bang Theory" fails to address the **fundamental question:** Where did all of the compact matter come from?

One **plausible answer** to this perplexing question is that the **matter and energy** that started the "Big Bang" has **always existed** in different forms throughout **eternity** and that this "Bang" was only the beginning of the **latest recycle** of cosmic matter and energy. Scientific evidence is now accumulating that would seem to support a **recycle** hypothesis.

However, there is also evidence that rather than an expected **slowing** in the rate of universe expansion since the "Big Bang" nearly 14 billion years ago, scientists have surprisingly discovered that distant galaxies are moving away from one another **faster** than ever. A new international program involving 23 scientific institutions is now investigating a mysterious cosmic propellant that scientists call **Dark Energy**.

Obviously, there are many puzzling questions yet to be answered satisfactorily.

*

Current **scientific evidence** appears to indicate that there are at least **nine basic principles** or natural laws operating in the formation, development and support of **life on planet Earth**. These basic principles include but certainly are not limited to the following:

- Principles of Genetics

- Principles of Opposites

- Principles of Randomness

- Principles of Competition

- Principles of Change

- Principles of Mathematics

- Principles of Molecules

- Principles of Reason

- Principles of Technology

Descriptions of Some Basic Principles

Principles of Genetics—

The most fundamental set of principles in the formation, continuation, and change of all the diverse and amazing life forms on Earth, both past and present, are the processes and mechanisms of **genetic inheritance.** Although humankind worldwide for countless generations has utilized the genetic effects of **selective breeding** and related techniques to produce a host of useful and interesting varieties of **plants and animals,** the **cause and mechanisms** for the changes taking place remained a **mystery.** It was not until the mid-19th century that an Austrian monk by the name of Gregor Mendel, studying the

inherited characteristics of different varieties of garden pea plants that he had crossed pollinated, began the human discovery and understanding of elements in the basic design of **genetic inheritance** pertaining to all life on Earth.

Progress moved rather slowly for decades resulting in the discovery of genes, chromosomes, and related genetic factors. The **big breakthrough** came about a century after Mendel's experiments. In 1952, James Watson and Francis Crick reported the startling discovery that **genes are arranged in ribbons of genetic substances** in different coded sequences wound together in the shape of the "double-helix." They were awarded the Nobel Prize for their tremendously significant findings that since have become basic to all scientific study involving genetic decoding and its applications.

These findings led to the ambitious international scientific effort—the **Human Genome Project**—to map all genes on the 23 pairs of human chromosomes and to sequence the 3.1 billion DNA base pairs that make up the chromosomes. Begun in 1990 with James D. Watson as its head and later headed by Francis S. Collins, the goal of the project was to enable scientists to understand

the double helix

the **basis of genetic diseases** and to gain insight into **human development**.

With the aid of newly developed, **super high-speed computers**, this monumental task was largely completed in 2000 when 85 percent of the **human genome** had been decoded, and ended in 2003 with 99 percent decoded.

Along the way, bitter **controversy** arose over attempts by commercial firms to **patent** specific human **genetic sequences**. Watson and Collins were **adamantly opposed** to genetic patenting and eventually **won the battle** contending that there should be **no ownership** of the "**laws of nature.**"

In order to study genetic similarities **among species**, the genomes of several organisms were also decoded during the same period of time. Surprisingly, the genome of the **mouse** was found to be very **similar** to that of the **human genome**, a finding that has proved to be very helpful in the search for treatments of specific **human diseases**.

*

The discovery of the principles related to the "double-helix" has led to understanding how these genetically-coded ribbons typically divide and combine to form new organisms very similar to their **parent organisms**. But **occasionally**, parts of the genetic ribbons can skip certain DNA codes and combine in ways that cause subtle changes in an organism. **Rarely**, however, parts of the genetic ribbons can even **break** and **combine** in ways that result in **major** differences in organisms. The scientific findings relating to the "double-helix", DNA, and genetic coding have opened the door to understanding the **basic design and operating principles** of genetic inheritance that over extended time have resulted in the myriad of **species and varieties** of life on planet Earth. The **principles of genetics** are truly fundamental parts of marvelous and magnificently functioning processes **perpetuating and changing** life on Earth.

Some persons, including Dr. Francis Collins the eminent Director of the National Institutes of Health, have termed DNA and genetic coding the "Language of God" and pointed out that amazingly DNA coding has only **four letters**—A, C, G, and T.

The Human Genome Project of the late 1990's mapped 3 billion base pair sequences that make up **human DNA**. In October 2010, scientists at Washington University School of Medicine and the University of Minnesota announced a five year project to diagram all major circuits in the **healthy human brain**. The task is staggering. The brain consists of 90 billion **neurons** connected by 150 trillion **synapses**. The researchers believe the project holds great promise to understanding **brain-based disorders**, such as alcoholism, autism, and schizophrenia.

*

Principles of Opposites—

Intertwined and closely related to the Principles of Genetic Reproduction are the **Principles of Opposites**. From the elementary study of **magnetism** comes the axiom that opposite forces **attract**, like forces **repel**. The basic principle of opposites attracting is so **pervasive** that it seems to apply generally throughout **nature** and the **Cosmos**, including gravity, magnetism, electricity, chemical composition, and atomic structure.

In the realm of **genetic reproduction**, the Principles of Opposites are fundamental to the mutual attraction of opposite genders. In higher forms of life, both animal and plant, the opposite genetic components of males and females combine in ways to produce offspring with characteristics closely similar to **one or both parents**, but **not always.**

It was this phenomenon of the effects of opposites in producing both similarities and differences in progeny that **Gregor Mendel** studied by cross pollinating garden peas when he began the long scientific search for discovery and understanding of the operating principles of **genetic inheritance**.

It should be noted again that long before Mendel's experiments, many people had applied in **practical and utilitarian ways** the principles of inherited characteristics in both plants and animals. Male parents having specific characteristics were often bred or crossed with females having other specific characteristics to produce **progeny** having more desirable characteristics than did **their parents**. This process of selective breeding when continued over many generations has resulted in both males and females that produce offspring with characteristics so **consistently similar** to their parents, yet so different from their distant ancestors or **original stock**, that they become a line or strain of **purebreds** or **thoroughbreds**, with pedigrees of ancestry. Through the centuries some persons have searched the environment to find cases where naturally occurring genetic crossings have propagated to develop different varieties of plants or animals that have **desirable or interesting** characteristics.

Although most people do not fret much over where or how the garden variety of **corn** they eat originated, among **plant experts** the history of corn (or maize) has been an intriguing **mystery** that generated much **controversy**. The argument bubbled and boiled for half a century. One camp held that corn's **ancestral** plant is a **weedy-type grass** called "teosinte" still found **growing wild** in remote areas of Mexico.

The opposing group of plant experts argued that teosinte could not have been the ancestor of modern corn and based their contention on old samples of **primitive corncobs** and **pollen** found in remote Mexican caves. This popular position held that corn's ancestor must have been a very primitive corn plant that had disappeared into **extinction**, never to be found, and thus **modern corn** was **not** the offspring of teosinte.

Recently, the science of **genetic sequence analysis** has shown clearly that **teosinte** is indeed the ancient forebear of modern corn. This scientific finding is important to today's **corn breeders** and others who seek to study corn's ancient ancestors in search of genes that might improve the **drought and disease resistance** of modern corn plants.

*

Principles of Randomness—

A third principle in the ongoing processes of life on planet Earth is the **Principle of Randomness**. The term **random** is often thought of as things done in a helter-skelter or disorderly manner. But among large populations of subjects or many events that interact over extended time, randomly occurring behavior and happenings take place with degrees of frequency that form patterns of outcomes in the **bell-shape** of the well-known **Normal Curve**.

The principles of the **normal probability** of randomly occurring events are fundamental to scientific research and statistical analysis of research findings and are also employed in many other aspects of human activity. For example, the same principles of random probability are involved in the manufacture of gambling devices, playing games of chance and considering the odds when placing a bet on a horse race or ball game.

The principles of **randomness** are important factors in the frequency with which inherited changes, both small and large, occur naturally and in the genetic propagation and distribution of plants and animals. The vast majorities of plant and animal progeny have characteristics very similar to their parents and tend to thrive in similar environments. Thus, they fall within the wide mid-range of the bell-shaped normal curve. See Figure 1.

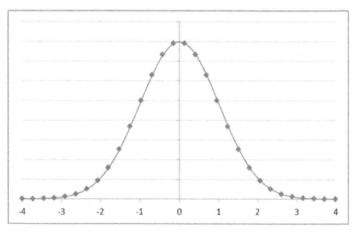

Figure 1 – Standard Normal Distribution

But some progeny have **small changes** or muta- tions in genetic characteristics from those of their **parents** (perhaps the result of inherited genes with some skipped genetic codes) that are not serious and fall **toward the ends** of the two-tailed **bell- shaped** curve. Some of these genetic mutant plants and animals are propagated and nurtured to become **new varieties** or strains of their plant or animal parents.

However, a **small number** of progeny have **major** changes in genetic characteristics so different from their parents (perhaps related to major breaks and recombining of genetic ribbons) that result in **severe** weaknesses that can lead to their death. These severely genetically damaged offspring fall **close to the ends** of the normal curve and are sometimes termed sports, malformed plants or crippled animals.

In contrast, a **tiny fraction** of progeny inherit **robust** genetic changes (again, perhaps related to major breaks and recombination of genetic ribbons) that enable them to **survive** and **propagate** in harsh conditions and events that tend to eliminate most or all of their peers.

*

Principles of Competition—

The vigorous nature of genetic reproduction for each species tends to produce an **overabundance** of its kind. Indeed, it is the vast overabundance of certain species, such as wheat, corn, rice, and, cattle, that supplies our grain, nuts, meat and other foods. Each **species** of animals and plants tends to expand over time to the maximum **limits** of its geographic and natural environment. Viewed individually, it would appear that a given **species** would soon spread unchecked to **dominate** its entire environment. But since there are typically many species present in the same environment, the basic **principle of competition** comes into play to maintain environmental balance.

Species **compete** with one another for **nourishment**, water, air, and where appropriate, proper amounts of **sunlight** or lack of light. Where **sufficient** amounts of these factors are present, species thrive or tend to **survive**. Where there are **insufficient** amounts of these factors, species become **extinct**. The **survival** of species can also be affected by predators, disease, catastrophe and climate change.

*

Principles of Change—

The **twin principles** of **change** and **motion** are fundamental and **pervasive** factors found to be operating throughout the universe. **Change** and **motion** are constantly observed among **celestial bodies** and the principles of change certainly apply to life on Earth.

Small changes in specific characteristics of certain varieties of **plants** and **animals** are often observed and possibly artificially manipulated many times in the span of a human **life time**, but **major** changes in plants and animals are **seldom** observed in such small time frames. Over **extended** time, however, there is much accumulated **evidence** from many sources that an amazing number of **major** changes in life forms have occurred in nature. Some changes culminated in **huge beasts** and plants that eventually became **extinct** while other much smaller organisms that showed surprisingly **little change** in structure persisted for countless **generations** and periods of time.

*

Principles of Mathematics—

Change and motion throughout the universe occur with amazingly precise applications of the highest **principles** of **mathematics**. It is **mathematical**

expressions of the predictable precisions of orbits, rotations, and distances of solar planets and stars that enables **space craft** from Earth to circle and land on planets, to fly by other planets, and to collide head on with a speeding asteroid. The **principles of mathematics** that describe the orbits and rotations of celestial bodies **also** relate to the genetic codes of DNA and to the structure of atoms. The principles of mathematics are truly integral and fundamental to the **basic operations** of the Cosmos and life on Earth.

*

Principles of Molecules—

Although atoms were long considered the smallest particles of matter, it is not until atoms are combined and linked to form **molecules** in chemical substances that atoms become intricately involved in the **processes of life**. The **principles of molecules**, therefore, are essential to life on Earth.

Molecules are the **smallest particles** of a substance that retain all the properties of that substance; consequently, molecules can be considered as the primary bases of all **living matter** and specifically for the chemical structure of **nucleic acids**, the essential substances for the building and genetic coding of DNA and genetic compositions.

The principles of molecular structure and chemical combinations are also at the heart of the vital processes of **photosynthesis** in which chlorophyll-containing tissue of plants use sunlight, carbon dioxide, and water to produce carbohydrates and oxygen, essential for sustaining nearly all forms of life. Thus, using **molecular principles**, the **chlorophyll** in plants captures **light energy** from the **Sun** and converts it into **chemical energy** in foods that sustain both **plant** and **animal life** on planet **Earth**.

*

Two Intriguing Principles of Creation and Change

The previous discussions relate to **seven Basic Principles** that can be considered as comprising some of the basic **structures** and **mechanisms** of **how** the Universe and life on Earth were **formed**, **operate**, and continue to **change**.

There are, however, at least two other sets of **Principles** pertaining to how **human intellectual perceptions** and **activities** have and are continuing to affect profoundly major conditions and changes in life on planet Earth. One is the **Principles of Reason**. The other is the **Principles of Technology**. These two sets of **Principles** deserve special attention and are discussed in this section.

Principles of Reason—

The **Principles of Reason** focus on the *perceived* specific **cause-and-effect** relationships between specific things, specific events, and specific related outcomes. Reasoning involves responses to either the *perceived* **causes**, to the *perceived* **effects**, or to **both**. These *perceived* causes and effects can be either **true** or **false**, or yet undetermined.

Most animals demonstrate use of varying types and degrees of basic **Reasoning** including, **cause-and-effect** reasoning, reward-and-punishment decisions, and possibly some degree of imagination.

As two animals **in the wild** cautiously approach each other in tall grass, the quickest and accurate response to questions— "Is it friend or foe?" …."Is it time for fight or flight?"—can decide the survival of the fittest.

In the **laboratory maze**, mice seem to use a form of trial-and-error reasoning in learning to find pathways to the food reward for their efforts. Well-trained guide dogs are remarkable in their judgment in helping sight-impaired persons move about safely. Sheepdogs display notable reasoning to the shepherd's command when herding sheep.

But animals generally lack the physical capacity of speech and manual dexterity. When the many factors or traits of basic **reasoning** are combined with the adroitness of human intelligence, speech and dexterity, a new phenomenon of **human power** over matter exists.

In human terms, **Reason** is the human faculty or power of acquiring knowledge that relates to **cause and effect** relationships, to determining, to assuming, or to believing the "why" or "how" specific things or events happen or do not happen. **Human reasoning** is much involved in establishing the existence and qualities of specific things and conditions, and in the important concepts of **right** and **wrong** attitudes, motivations, and behaviors that become the standards of **morality** and **behavior** among individuals, groups, cultures, and societies.

Among human beings, questions of the actual **truth** or **validity** of the existence of specific events, conditions, or things are often difficult to answer to the general agreement of others. Thus, these difficulties have often led and frequently do lead to disagreements, quarrels, lawsuits, feuds, schisms, and warfare.

The French play, *Chantecler* by Edmond Rostand, vividly satirizes how much **trouble** and **hostility** was created among his peers by *Chantecler* the rooster's **false**, but sincere and ardent **belief** and zealous **assertion** that it was his **crowing** that caused the **Sun** to rise.

The **Druids** of ancient Britannia held sacred the **Mistletoe,** the parasitic plant that grew on large oak trees. They **believed** the small bunches of evergreen **Mistletoe** leaves high in the winter leafless oak trees preserved the trees' life substance during winter, and come spring, restored life to the trees.

The ancient **Egyptians** held the **Ibis bird** sacred. Each spring when the Nile River flooded the bottom lands and deposited rich silt soil so vital to all Egypt, whole flocks of long legged, long-necked **Ibis birds** also showed up wading around in the marshes looking for food. The Egyptians **believed** the **Ibis birds** had brought the floods with them. This cause-and-effect **belief** that the annual arrival of the **Ibis birds** caused flooding of the Nile was so important to the Egyptians that they represented **Thoth,** their god of **wisdom,** learning and magic, as a man with the **head** of an **Ibis bird.**

It should be noted that there is an almost perfect **correlation** (perhaps above 0.98) between the two actions happening together in **each** of the three cases described above—the **crowing** of rooster and **rising** of Sun; the winter-green **Mistletoe** and **spring life** of oak trees; the **annual** arrival of **Ibis birds** and **flooding** of the Nile river.

But we should keep in mind that close **correlation** of two factors or events **does not** indicate a **cause-and-effect** relationship between them. Statistical correlation data only **indicate** the **degree** to which two factors **happen together**—nothing more—and not that one factor **affects or responds** to the other.

Most ancient pagan **pantheons of gods** were based on the sincere and ardent belief that certain gods **caused** specific things **to happen** or not happen and that specific **worship**, devotional **rituals**, or **sacrificial** practices would **appease** a specific god's wrath or **court** a specific god's **favor**. Much of the lives and activities of individuals and groups of ancient peoples centered on **appeasing** one god's **wrath** and **courting** another god's **favor** and **protection**.

Sincerity, ardency, and zeal do not establish the **truth** or **validity** of a **belief** that certain things **cause** other things to **happen** or not happen. Thus, the separation of **facts** from **non-facts** is an abiding and important concern in the process of **reasoning**.

*

Critical Thinking, **Logical Thinking**, and **Logical Argument** are closely related components of **human reasoning**. These methods of thinking are the purposeful, orderly, and reflective **judgments** about what to believe is **true** or what to **do** in response to observations, experiences, evidence, information, or arguments.

Varying aspects and degrees of **logical thinking** and **logical argument** have been used in the **development** and **maintenance** of tribal cultures and civilized norms from the beginnings of human kind.

Various forms and rules for **logical argument** have been used in **tribal councils** and complex **civil trials** to decide whether there is **sufficient** evidence to find persons **innocent** or **guilty** of specific crimes or offences.

Forms of **logical argument** were used by both the prosecution and the defense in ecclesiastical **tribunals** such as the trial of **Galileo** Galilei for **heresy** in 1633. Galileo was charged with **advocating** and **promoting** the **radical** Copernican heliocentric theory that the **Sun** is the central body of the **solar** system and that the Earth and planets **revolve** around the Sun and that the Earth **rotates** daily on its axis.

Galileo was **found guilty** of erroneous and dangerous **heresy** of church **dogma** that the **Earth** was the central body of the **solar** system and the **Sun**, the **planets**, and the **stars** revolve around the **Earth**. Following the **guilty verdict of heresy**, Galileo was threatened with a **sentence of torture** if he did not **recant** of his beliefs and teachings. To avoid torture, he did **recount**, a lighter sentenced was imposed and he spent the rest of his life under house **arrest**.

As a famous scientist once said, "Smart people (like smart lawyers) can come up with **very good** explanations for **mistaken** points of view."

*

Within the past three centuries, the process of **critical thinking** has been developed into widely recognized and extensively used **scientific methods** of sorting out **truth** from **untruth**.

Scientific methods are the processes by which scientists, collectively and over time endeavor to **construct accurate**, reliable, consistent, and non-arbitrary representations of specific **things** and **events**. Recognizing that personal and cultural beliefs influence both perceptions and interpretations of observable phenomena, scientists aim through the use of **standard scientific procedures** and criteria, including structured and repeated observations, to minimize those influences when testing an hypothesis or developing a theory. Scientists wish that their findings and conclusions will be verified and confirmed by other respected scientists but are ready to accept evidence that their findings and conclusions are not supported.

*

Learning and **Imagining** are two important aspects of human Reasoning. **Learning** is the storing and recall of knowledge or behavior that is either newly found by the learner or that is acquired from the findings, assumptions, or beliefs of others. **Imagining** is the forming of mental images of something or relationships that are not actually present to the senses. **Imagining** is to picture, to hear, to envision, to suppose, to guess, to hypothesize, or to postulate an existence of specific things or relationships among various things.

It is **imagining** that creates anxiety in some persons as they enter a reportedly "haunted" house on a dark night, or walk past a dark and creepy cemetery, or read a scary ghost story.

It is **imagination** that delights eager children as they hear the poem *The Night before Christmas* read to them on Christmas Eve. It is **imagination** that engages and interests thoughtful adults as they read a good story or novel.

It is also **imagination** that envisions a team of scientists, mathematicians, and engineers as they earnestly consider the possibilities and techniques needed to accomplish their mission to land a man first time on the Moon and return him safely to Earth.

The **Principles of Reason** and the related processes have and continue to affect greatly how humans and other beings live and act on planet Earth.

*

It should be noted in passing: The readers of this Essay are asked to apply careful logic and critical thinking when considering this Essay's basic **premise**, the firm **commitment**, and the **evidence** presented about **how** the universe and life on Earth were created and are being sustained.

*

Principles of Technology—

Technology is the product of human **reasoning** when used to supply the needs and wants of human beings. **Technology** includes the many ways in which people both provide and use the tools, objects, possessions, conditions, concepts, knowledge, and expertise needed or desired for themselves, their groups, their cultures, or their civilizations.

Similarly, **Technology** is also the product of perhaps a combination of both animal reasoning and instinct used to supply the needs and wants of specific animals and even some insects. Consider the beaver—deciding where to build a dam to form a small lake in which to build a home, selecting and preparing building materials, design and construction of beaver dam and home—a complicated construction job. Also consider birds building nests, foxes building dens, bees constructing combs and filling them with honey.

*

The bounds of **human technology** appear to be limited only by human imagination and ingenuity of what is possible with determined and persistent application of valid cause and effect relationships.

In 1780, Benjamin Franklin predicted:

> It is impossible to imagine the height to which may be carried, in a thousand years, the power of man over matter. We may perhaps learn to deprive large masses of their gravity, and give them absolute levity, for the sake of easy transport. Agriculture may diminish its labor and double its produce; all diseases may by sure means be prevented or cured, not excepting that of old age, and our lives lengthened at pleasure even beyond the antediluvian standard.

Within 233 years, most of these predictions have been exceeded and the others are well on the way toward accomplishment.

Technology is **accumulative, compounding**, and **accelerating**.

Within 123 years after Franklin's prediction of humans learning how *to deprive masses of their gravity*, the Wright Brothers flew the first engine powered airplane 120 feet in 12 seconds at Kitty Hawk, NC.

In 24 more years, Charles Lindberg flew nonstop New York to Paris, 3500 miles in 33 and one-half hours.

In 42 more years, Neil Armstrong and Buzz Alden were walking on the moon, planting an American flag, installing scientific equipment, picking up samples of Moon rocks, and were returned safely to Earth with the collection of Moon rocks.

In 35 more years, an Earth-made robotic machine was roving around the surface of planet Mars performing extensive geological analysis of Martian rocks and planetary surface features and continued to do so for more than 7 years, viewed and controlled by scientists on planet Earth.

And within 218 years of Franklin's 1,000 year predictions of humankind's learning how *to deprive masses of their gravity,* the first modules of the International Space Station were being launched to become a human- habitable structure for scientific space exploration in permanent orbit around the planet Earth.

This was all accomplished by **Technology** that was and is focused on:

- Harnessing and controlling various forms of energy to perform specific tasks and func- tions desired by human beings.

- Developing concepts, values, and relation-ships about how persons can and do live and work together in groups and civilizations.

- Accumulating, sharing, and transfering human knowledge.

- Constructs which hold that things, conditions, concepts, and scientific theories are not final or absolute but always open to possibilities for change and improvement.

- Determining what is fact and what is fiction.

- Determining what is workable and what is unworkable?

However, these and many other amazing and useful technologies also have a dark side of incredible and horrible misuse.

In the little over two centuries since Benjamin Franklin's insightful predictions, humankind has been plagued with periodic warfare with each war using more deadly and destructive weaponry than the previous and in the process killing millions of people, wasting vast resources, and accumulating staggering debts.

*

During World War II, new pilotless "buzz bombs" (V-1 type flying bombs) rained hundreds of tons of high explosives on targeted cities in Britain many scores of miles away killing and wounding many hundreds of people. At the same time, many sorties of huge super bombers conducted saturation and "carpet-bombing" of distant German cities killing many thousands of people.

The American/Atlantic Theater of WWII ended May 8, 1945 with the **unconditional surrender** of the German Army to Allied forces. But the fighting in the American/Pacific Theater raged on with increasing ferociousness as Allied Forces moved ever closer to the Japanese Homeland.

In July 1945, the victors of the **unconditional surrender** of the German Nazi regime met in Potsdam, Germany to settle the aftermath of the War in Europe and to end the War with the Empire of Japan.

On July 26, the conference issued a declaration giving Japan the ultimatum of **unconditional surrender** in the name of the United States, Great Britain and China (Soviet Union was not then at war with Japan). or meet "prompt and utter destruction." Japan **refused** to surrender.

Then came the big one! On August 6, 1945, an astounding new technology that enabled a **single** super-fortress bomber, carrying a **single** mega explosive **Atomic bomb** (code named "Little Boy") to destroy the entire city of Hiroshima, Japan, immediately killing between 90,000 and 100,000 people and fatally wounding by blast and radiation unknown thousands of people.

Within 16 hours after the A-Bomb exploded, U.S. President Harry Truman issued a Press Release explaining to the American people and the World what had happened and saying to the Japanese, surrender or the bombings will continue. The US President declared:

> We are now prepared to obliterate more rapidly and completely every productive enterprise the Japanese have above ground in any city. We shall destroy their docks, their factories, and their communications. Let there be no mistake; we shall completely destroy Japan's power to make war.

The next day (August 7), US planes dropped warning pamphlets on many Japanese cities.

On August 9, 1945, a second **Atomic bomb** (code named "Fat Man") was dropped on Nagasaki, Japan, with similar devastation and loss of life as Hiroshima.

At noon on August 15, 1945, for the first time in history, the voice of the Japanese emperor was heard on the radio. His Imperial Majesty Hirohito had recorded the message the previous day and spoke to his people in the very formal language of the Imperial Court. The Emperor declared:

> We have ordered Our Government to communicate to the Governments of the United States, Great Britain, China and the Soviet Union that Our Empire accepts the provisions of their Joint Declaration.

This was **unconditional surrender!** The war was ended saving untold hundreds of thousands of human lives—military and civilians plus animals, property and resources.

News of the awesome and unbelievably devastating explosive power of Atomic Bombs set off a frantic race between the World's two military super powers to manufacture enough nuclear bombs to destroy the other. Competition was fierce to develop ever more destructive nuclear fission Atom Bombs. Soon came the unbelievable mega-destructive invention—the nuclear fusion Hydrogen Bomb that releases the power that fuels the Sun! A single nuclear fusion Hydrogen Bomb releases as much as 25,000 times the yield of the nuclear fission Atomic Bombs dropped on Hiroshima and Nagasaki, Japan.

Not to be left behind, other countries began producing their own stock piles of nuclear weapons. And now the bomb-making nations of the World have stockpiles of far more than enough nuclear weapons to virtually destroy all civilization on Earth.

Fortunately, apparently for fear of mutual destruction, no nation has yet ventured to use a nuclear bomb against another nation.

*

But let us return now to the many bright and positive ways that technology is changing how people live, work, and play peacefully, productively, and enjoyably on planet Earth.

Consider the many networks of interacting technologies that make possible the following amenities in these pleasurable circumstances.

A man and wife live in a comfortable, air-conditioned apartment on an upper floor of a Boston high-rise that overlooks the city and bay. The couple has invited a few friends for dinner and to watch, on their new 70 inch LED screen 1080p 120Hz HD TV (made in China) a championship football game played in California.

The dinner menu includes: Appetizer fruit cup with grapes from Chile, honeydew from Guatemala, Kiwi fruit from New Zealand, and pineapple from Hawaii. The entrée is a choice of Argentine beef steak or baked Alaskan Salmon. The vegetables are Idaho scalloped potatoes and steamed spinach from Florida. The dinner wine is from France. The dessert is chocolate cake from Germany topped with Ben and

Jerry vanilla ice cream from Vermont. The coffee was from Colombia.

But the dinner party would not be possible in this lovely and comfortable setting, if a few years before, skilled operators using multi-story crane technology had not lifted the huge steel girders many stories above ground and guided them carefully into position for bolting and making ready for the next level girders.

Moreover, the dinner party would not be possible in this pleasurable location without the swift and dependable elevator technology that raised and lowered the party guests confidently to their desired floor with the mere press of a button.

*

Human curiosity creates some of the most pressing demands for new technology and how to use it. Curiosity is the human desire to know—What's out there, What's in there, What's over there, How does this thing work, etc.?

Galileo's invention of his telescope in 1609 started a chain of activities to answer human curiosity about "What's out there in space?" Galileo found the Moon was not made of bright green cheese or other strange things that some in his day speculated, rather the Moon was made of rocks and craters and stuff similar to that on Earth. Then came **curiosity** about the red planet Mars. Telescopic views of the planet seemed to show there were canals on Mars. Could that mean there is life on Mars? Bigger telescopes revealed no signs of life and that the red planet was apparently mostly dry.

Benjamin Franklin could not have imagined the amazing things that would happen when human beings, as he had predicted, learned to deprive large masses of their gravity and to give them absolute levity. The technologies of putting telescopes in orbit high above Earth's thick protective atmosphere did just that—it deprived the telescopes of their gravity and gave them absolute levity. This enabled space telescopes to collect pictures and data unhindered by atmospheric twinkling and other interferences and enabled the space telescopes to capture amazing images of cosmic events which drastically changed previous concepts of the cosmos, space, and time.

*

In 1990, a Space Shuttle carried the Hubble Space Telescope into Earth orbit. It was not the first space telescope to be put in Earth orbit, but Hubble was the largest and the only space telescope designed to be serviced by astronauts in space. It was also equipped to take full-color images using the full-spectrum of visible light plus equipment that captured different other types of electromagnetic radiation.

When in orbit, the first images taken by Hubble appeared hazy and dull—a small flaw had occurred in grinding and polishing the telescopic mirrors. A team of astronauts aboard a Space Shuttle installed what in effect was a pair of corrective glasses. The glasses worked great! The Hubble telescope could see clearly and began sending fantastic color images of galaxies of stars and other amazing things millions of light-years away from tiny planet Earth.

Public and professional media sources immediately began beaming these amazing images to people all around the world, creating widespread wonder and curiosity about," What's **really** out there?" This boosted public pride in the U.S. National Aeronautic and Space Administration (NASA) and increased public support of space discovery and exploration. Similar excitement and actions occurred among the public and scientists in many other countries. Teams of astrophysicists and other scientists around the world shared their findings and joined in space discovery and exploration.

A number of space telescopes now in Earth orbit were designed specially to capture images and collect astrophysical data using specific frequencies of the vast range of frequencies in the **Electromagnetic Spectrum**. The Electromagnetic Spectrum of a **specific** object is the characteristic distribution of electromagnetic radiation emitted or absorbed by that object.

Space telescopes grouped by frequency ranges include: Gamma rays, X-rays, Ultraviolet rays, Visible light rays, Infrared light rays, Microwave rays, and Radiowave rays.

In addition, there are space telescopes that collect astroparticles such as Cosmic rays (atomic nuclei and electrons) plus those that measure Gravitational waves.

Space telescopes continue to send millions of images and reams of scientific data about events and things hundreds of millions of light years away from Earth. For more than two decades astrophysicists and other space scientists have been poring over these cosmic images and data findings and verifying evidence of amazing things and astounding events.

They have found within the Earth's Milky Way galaxy evidence of "**Black Holes**" where the remains of dead stars are apparently being pulled in, disassembled and "**star-parts**" prepared for recycling. Space-scientists have also found evidence of "Stellar Nurseries" where the formation of new stars is taking place using **stellar remnants** and a diffuse **interstellar medium** (ISM) of gas and dust consisting roughly of 70 percent **hydrogen** (by mass) and the other 30 percent consisting mostly of **helium** and other gases An example of **Stellar Nurseries** is shown in the widely publicized Hubble telescopic image known as *Pillars of Creation* where stars are forming in the Eagle Nebula.

In 2011, three teams of scientists—at Notre Dame in South Bend, the Space Telescope Science Institute in Baltimore, and the University of Massachusetts in Amherst–using the Cosmic Origins Spectrograph abroad the Hubble Space Telescope identified vast "halos" of gas **surrounding** a selection of **40 galaxies.** Previously invisible, these halos were much larger than previously believed and apparently act as reservoirs for what the scientist termed recycled star "stuff"—light elements such as hydrogen and helium and heaver elements such as carbon, oxygen, nitrogen, and neon.

Space-scientists and other scientists are also finding answers to two long-asked and intriguing questions.

Question 1: Do any stars have planets revolving around them like our Sun has? **Answer**: Definitely Yes. They are called Host Stars and the planets around them are called Exoplanets. In 2009, NASA placed in orbit the Kepler spacecraft observatory with the specific mission to discover Earth-like planets orbiting other stars. NASA keeps a running record and archives of verified Exoplanetary discoveries. The current and fast growing count is 693 Exoplanets around 552 Host Stars, plus 1,235 Planetary Candidates.

Question 2: Do any planets have climates that could support life?

Answer: Probably Yes. Scientists are now honing in on definite answers to the question. They have developed *The Habitable Exoplanets Catalog* that includes the *Periodic Table of Exoplanets* that divides exoplanets into six mass and three temperature groups (18 categories total). As exoplanets are discovered and verified, each is placed in its appropriate mass/temperature category. Exoplanets that fall near the center of the Table (termed by some the "Goldielocks " zone) are most likely to have life-supporting climates. As of February 2012, four exoplanets had been found in the possible life-supportable group

Persons might ask, "How do scientists know all this?" The answer is a cosmic version of the old saying, "It's like looking for needles in a hay stack." In March 2009, as cited above, NASA launched the **Kepler** spacecraft observatory (named in honor of Johannes Kepler, the 17th century German astronomer) with the mission to discover Earth-like planets orbiting other stars in the Earth's portion of the Milky Way galaxy and to determine how many of the billions of stars in the galaxy potentially have habitual exoplanets.

Kepler's only instrument is a photometer that continuously monitors the brightness of over 145,000 main stars in a fixed field of view. The data are analyzed to detect periodic dimming caused by exoplanets that cross in front of their host stars. The data are further analyzed to detect if the exoplanet has possibilities of supporting life.

A recent close-up example of this process occurred on June 5, 2012, when the "exoplanet" Venus transited across the disc of its Host Star (our Sun) and "exoplanet" Earth—a widely NASA photographed and published event that will not be visible again for 105 years.

*

Human curiosity expands technology in two opposite directions: One, to answer: "What's out there in space?" The other, to answer: "What's in these **tiny** things all around us on Earth?" Interestingly, initial investigations in both directions, sky and ground, were based on the same technology of **glass lens** magnification—**Telescopes** and **Microscopes**. Also, advanced research instruments in both directions use **similar bands** of the huge **Electromagnetic Spectrum**.

In 1674, Antonie van Leeuwenhoek, a Dutch lens maker, built the first practical microscope to examine blood cells, yeast, insects, and other tiny objects. Leeuwenhoek is famous for his discoveries of bacteria and various single cell organisms. These discoveries started the chain of activities to answer human curiosity about "What's in such tiny things?"

In 1830, Joseph Jackson Lister greatly improved the single lens microscope by showing the distortions caused by the spherical single lens could be eliminated by several weak lenses used together at certain distances to give good magnification without blurring the image. This was the prototype for the compound microscope.

In 1903 Richard Zigmondy developed the ultramicroscope that could study objects below the wavelength of white light. He won the Nobel Prize in Chemistry in 1925.

In 1931 Ernst Ruska invented the electron microscope that depends on **electrons** rather than light to view an object, electrons are speeded up in a vacuum until their wavelength is extremely short, only one hundred-thousandth that of electromagnetic wave length of white light. Electron microscopes make it possible to view objects as small as the diameter of an **Atom**.

The ability to view Atoms, long considered the smallest unit of matter, created much curiosity about "What are Atoms made of?" and generated pressure to develop technology to learn the composition of Atoms and to find ways to use the tremendous energy of Atoms.

In 1934, Ernest Lawrence at the University of California, Berkeley invented the Cyclotron to study the nuclear structure of the atom. The cyclotron produced high energy particles that were accelerated outwards in a **spiral** rather than through an extremely long, **linear** accelerator. Advanced variations of this technology are now used extensively in home and business appliances including: Computers, TVs, and Microwave ovens.

In scientific research, **Super Collider Particle Accelerators** are now being used to study subatomic parts of Atoms. These Super Colliders, often called "Atom Smashers", accelerate oppositely charged subatomic particles, such as atomic nuclei, in opposite directions at speeds approaching that of light and crash them head-on, breaking them into still smaller pieces to learn more about the composition of Atoms. The subatomic particles include: Quarks, Anti-quarks, Protons, Neutrons, Baryons, Bosons, Mesons, Atomic Nuclei, etc.

The largest Super Collider Particle Accelerator ever built, called the Large Hadron Collider (LHC), is the new 17 mile underground circular accelerator constructed by the European Organization for Nuclear Research (CERN) crossing the Swiss/French border is now being tested. A primary purpose of this massive undertaking is to find evidence of the theoretical Higgs boson subatomic particles that quantum physicists theorize are the ultimate particles that add mass and shape to all Atoms on Earth and in the Universe.

There is much excitement in the scientific community and in the public media about the search for the Higgs boson particle termed by Noble Prizewinning physicist Leon Lederman and relished by the popular media as "The God Particle."

The July 5, 2012 *Wall Street Journal* carried a half-page news item headlined:

> *Discovery May Help Tell Universe's Secrets: After Half-Century Search, Scientists Pin Down Higgs-Like Particle, Closing In on Explanation for Why All Objects Exist.*
>
> Scientists said they found a subatomic particle resembling the long elusive Higgs boson, a landmark discovery that could explain why particles have mass, and by extension, explain why stars, planets and other objects in the universe exist at all.
>
> On Wednesday morning, hundreds of scientist assembled at the European laboratory CERN in Geneva and many tuned in to a live webcast to hear how fresh data from the large Hadron collider had conclusively revealed the existence of a Higgs-like particle.

"We have observed a new boson," said Joe Incondela, of the University of California, Santa Barbara, a member of the group reporting the new data....

The particle's namesake, British physicist Peter Higgs...one of several physicists in the 1960s to predict its existence [said] "It is an incredible thing that has happened in my lifetime."

*

Micro-scientists, using **microscopic-technology** to study small subatomic parts of Atoms on Earth, are finding evidence of the same kinds of particles and electromagnetic waves that **space-scientists**, using **telescopic-technology**, are finding among dying stars swirling around cosmic **Blackholes** being pulled in apparently for disassembly and their "star-parts" prepared for recycling. These same type of "star-parts", subatomic particles, and electromagnetic radiation appear near and among "Star Nurseries" or "Pillars of Creation" in our own Milky Way galaxy and perhaps in other galaxies of the cosmos.

*

Here on Earth, a group of **Enabling Technologies** deserve special recognition for the multiversity of technologies they provide, serve, and expedite. These Enabling Technologies include: **Computer**—rapid data and image processing, storage and access; **Radio**—rapid audio transmission and reception; **Television**— rapid video and audio transmission and reception; **Digital Photography**—rapid image capture,processing and storage,and **Internet Related** technologies.There are other Enabling Technologies, but these are examples of the important accumulative and compounding aspects of technology.

*

As amazing and wonderful as these tremendous technologies would seem, their existence and continued development depend upon two interrelated technologies essential to human **Civilization**:—one is the technologies of civil **Governance**, and the other is the technologies of effective **Economics**.

The technologies of civil **Governance** with the exercise of authority, control and focus of human activities are basic in determining the **directions** and **ways** people live and work, presently and in the future.

The technologies of effective **Economics** deal with the production,distribution and consumption of goods and services that sustain human welfare and provide rewards for human labor, human thought and human action that are basic to the output and usages of most technologies.

*

Michio Kaku, a renowned quantum physics professor at City University of New York describes his position in these words:

> I am a quantum physicist. Every day I grapple with equations that govern the subatomic particles out of which the universe is created. The world I live in is the universe of eleven-dimensional hyperspace, black holes, and gateways to the multiversity. But the equations of the quantum theory, used to describe exploding stars and the Big Bang, can also be used to decipher the outlines of the future.

Professor Kaku interviewed over 300 of the world's outstanding scientists, who are now inventing the technologies of the future in their labs, to get their best thinking about the future of technology in the year 2100. Note, this is less than one-tenth the 1,000 years Benjamin Franklin used in his remarkable 1780 predictions.

Kaku compiled his and the thoughts of his fellow scientists about the future, in the book he published in 2011, titled *Physics of the Future: How Science Will Shape Human Destiny and Our Daily Lives by the Year 2100.*

Michio Kaku and the scientists he consulted view future technology developments with enthusiasm and with the expectation of what the scientist term a "Planetary Civilization" by the year 2100.

The term "**Planetary Civilization**" applies to what physicists term a "**Type I Civilization**"which is considered to be the ultimate level of civilized **technology** to use efficiently the **total energy** available on Earth and coming to planet Earth from the Sun. Professor Kaku and fellow scientists calculate that currently global **technology** is only about **70 percent of the way** toward reaching the **full potential** use of global energy on a sustainable scale.

Michio Kaku and the scientists express confidence that the accumulating technology will reach the goal of total utilization of Earth's sustainable energy needed for a **Planetary Civilization** by the year 2100.

What an intriguing possibility—to live and participate in the global development of a Planetary Civilization where the grandchildren of the present generation have the realistic prospects of living, participating and enjoying life in a Planetary Civilization environment with bountiful and continuing sources of energy.

Such realistic prospects open whole vistas of possibilities of how the present earthly civilization might change with plentiful supplies of petroleum-type energy, natural gas, electrical energy, atomic and sub-atomic energy, abundant fresh water, sufficient supplies of today's scarce minerals, stronger-than-steel spider-silk fibers, and other amazing technologies yet to come.

Before dismissing these possibilities and projections as far-off, Star-Trek, futuristic, space fantasy, consider the following recent news items.

On Oct 24, 2012, the AOL, Auto Staff, posted an update on AOL of the current status of World energy for Auto-interested readers that carried this surprising headline:

Top Oil-Producing Country Intends to Focus '100 Percent' On Renewable Energy.

Prince says move 'very good for the world'

Saudi Arabia is the world's top producer of oil, extracting approximately 11.6 million barrels every day. The oil takes care of approximately two-thirds of the kingdom's own energy needs and is the lynchpin of the country's lucrative exports.

So how is the oil-rich country planning for its energy needs in the future? By focusing on renewable energy.

Earlier this week, Prince Turki Al Faisal Al Saud, a top spokesperson for Saudi Arabia, said that Saudi Arabia intends to generate 100 percent of its power from renewable sources, such as nuclear, solar, and low-carbon energies.

"Oil is more precious for us underground than as a fuel source," said the prince, whose country holds approximately 20 percent of the world's oil reserves, according to the International Energy Agency. "If we can get to the point where we can replace fossil fuels and use oil to produce other products that are useful, that would be very good for the world."

In related news, the U.S. is giving Saudi Arabia a run for the title of world's top oil-producing country, a boom that has left many experts surprised....

"Five years ago, if I or anyone had predicted today's production growth, **people would have thought we were crazy**," Jim Burkhard, the head of oil markets research at IHS CERA tells The Huffington Post.

The Wall Street Journal (April 25, 2012) had a headline that read: **Asteroid-Mining Venture Sees $100 Billion Prize.** The article was accompanied by a computer-generated image showing a "rendering of a spacecraft preparing to capture a water-rich, near-Earth asteroid."The lead paragraphs stated:

> Seattle—A start-up with high profile backers on Tuesday unveiled its plan to send robotic spacecraft to remotely mine asteroids, a highly ambitious effort aimed at opening up a new frontier in space exploration.

> At an event at the Seattle Museum of Flight, a group that included former Aeronautic and Space Administration officials unveiled Planetary Resources, Inc. and said it is developing a "low-cost " series of spacecraft to prospect and mine "near-Earth" asteroids for water and metals, and thus bring "the natural resources of space within humanity's economic sphere of influence."

> The solar system is "full of Resources, and that we can bring back to humanity," said Planetary Resources co-founder Peter Diamandis, who helped start the X-Prize competition to spur non-governmental space flight.

> The company said it expects to launch its first space craft to low-Earth orbit—between 100 and 1,000 miles above Earth's surface—within two years, in what would be a prelude to sending spacecraft to prospect and mine asteroids.

About a month later on May 23, 2012, the Associated Press carried an article headlined: **Commercial space race gets crowded behind SpaceX.**

> WASHINGTON (AP) — A privately built space capsule that's zipping its way to the International Space Station has also launched something else: A new for-profit space race.
>
> The capsule called Dragon was due to arrive near the space station for tests early Thursday and dock on Friday with its load of supplies. Space Exploration Technologies Corp.— run by PayPal co-founder Elon Musk — was hired by NASA to deliver cargo and eventually astronauts to the orbital outpost.
>
> And the space agency is hiring others, too.
>
> Several firms think they can make money in space and are close enough to Musk's company to practically surf in his spaceship's rocket-fueled wake. There are now more companies looking to make money in orbit — at least eight — than major U.S. airlines still flying.....
>
> NASA has given seed money and contracts to several companies to push them on their way. But eventually, space missions could launch, dock to a private space station

or hotel and return to Earth and not have anything to do with NASA or any other country's space agency.

On June 1, 2012, The Wall Street Journal reported, "SpaceX Splashdown Goes Smoothly: Dragon Capsule Returns From Space Station, Marking Milestone for Commercial Space Enterprise."

Los Angeles—The first private spacecraft to visit the International Space Station made a dramatic return home with a precise splashdown in the Pacific Ocean off the Southern California coast....

The capsule performed nearly flawlessly throughout and preliminary estimates indicate that it landed within a mile of its target... It was loaded with roughly, 1,400 pounds of old equipment, returning scientific material and other cargo.

*

It should be noted, Professor Kaku observes that progress toward the eventual destination of the long voyage into the realm of science and technology is resulting in many of the changes and upheaval around us today. Kaku concludes:

This transition is perhaps the greatest transition in history, marking a sharp departure from all civilizations of the past. **Every**

headline that dominates the news reflects, in some way, the birth pangs of this planetary civilization. Commerce, trade, culture, language, entertainment, leisure activities, and even war are being revolutionized by the emergence of this planetary civilization. [Emphasis added]

However, **technology,** as observed previously, is accumulative, compounding, and accelerating. Considering the speed with which technologies accelerated in the past half-century, compared with the current accelerating speeds that the **Globalization** of major areas of technology are occurring now including — globalization of scientific research, manufacturing, communications, transportation, energy production and utilization, agricultural production, human health quality, human welfare quality, and in the technologies of civil **governance** and effective **economics**—it seems that Professor Kaku and the present day scientists might have been a bit overly conservative in projecting almost a century for a **Planetary Civilization** to become a reality on planet Earth.

It should be remembered, though, that much depends on how these tremendous and powerful technologies are used. Intelligent and proper use could lead toward a benevolent utopian-type civilization. Severe misuse of these tremendous and powerful technologies could result in setting human civilized progress back a half-millennium years, or perhaps result in the complete annihilation of civilization as we know it.

Truly, the **Principles of Reason** and the **Principles of Technology** are two of the basic sets of principles or laws of nature involved in creating, sustaining, and continuing to change the Universe and life on planet Earth.

*

Summary and Conclusions

Controversy has long raged over one of the most profound questions of our time. *Which of two ways was the Universe and all life on Earth created and continues to operate?* One is called Creationism and the other called Evolution.

Creationism is the theology-based belief that the Universe and all Life on Earth was created by God in the brief time of literally one, six-day work week by use of Divine commands and then a seventh day of Divine rest.

Evolution is the science-based theory that creation of the Universe and all life on Earth took place, is being sustained and changed over extended time by natural laws and principles of nature. (or sometimes called Darwinism.)

Unfortunately, even in this current DNA-Space-Age Era, arguments regarding these issues are much the same as those used a century ago. There is need for a fresh and current examination of the information now available that pertains to these issues.

This Essay is a sincere attempt to present fairly, justly and objectively past and current evidence pertaining to **both ways** that could have been used in creation, so that readers who are truly interested in learning **how** the Cosmos and all life on planet Earth were created and continue to operate can decide for themselves how creation was actually accomplished.

To find answers to the important question: *How was the Universe and all life on Earth created and sustained,* we began with the basic premise that the Cosmos and all life on Earth do now exist and in some way they were all created. We also began with the firm commitment to learn as much as we could about **how** all that is in nature and the Universe was created

Two basic sources of information about **how** the Universe and life on Earth were created were examined for **evidence** of the ways and means used in the process of creation: 1) the *Holy Bible* and 2) the findings of scientific research.

The first 34 verses of *The First Book of Moses* titled *Genesis* in the *Holy Bible* were found to contain the only descriptive evidence in the *Bible* of **how** God created the heavens and the Earth in six days and completed the work on the sixth day with the creation of humans, "male and female he created them."

God blessed them and told them to be fruitful and multiply."And God saw everything he had made, and behold, it was very good.""God blessed the seventh day and hallowed it, because on it God rested from all his work which he had done in creation."

It was noted that in this six-day account God created the day and night on the first day, plants and trees were created the third day, but the Sun and Moon were not created until the fourth day.

These first 34 verses of *Genesis* can be viewed either as the **revelation** from God to mankind about **how** the heavens and Earth were created in six days; or viewed as the result of human efforts to **discover** and understand **how** God created the heavens and Earth based on observations of the things and events persons perceived about them and around them, unaided by telescopes, microscopes, or other technologies.

Regardless of **revelation** or **discovery**, these 34 verses of *Genesis* deserve recognition for the tremendously important role they have had in satisfying the concerns of millions of people through numerous centuries about the origins of the heavens and life on Earth and in the shaping of Jewish and Christian religious thought and culture. Such contributions include our seven day-week calendar, the religious Sabbath, and the six-day workweek even in societies that practiced slavery.

It is important to remember that many members and groups of both Christians and Jews no longer believe in the **literal** interpretation that God created the Universe and life on Earth **literally** in one six-day workweek and many now believe that the Universe and all life on Earth were created over extended time by natural processes.

Over the past two centuries, there have been concerted efforts and much progress in the scientific study of natural phenomena. Scientists have been systematically investigating and studying various events, conditions, circumstances, occurrences, and experiences using scientific methods of objective thinking and exploration to learn true cause and effect relationships of natural happenings and conditions.

In essence, **scientists** can be viewed as using their human intellect and rational thought to **discover** and **understand** various parts and pieces of the structure of the Universe and life on Earth.

Current scientific evidence indicates that there are at least nine basic principles or natural laws operating in the development and support of life on Earth. There are certainly many others, but nine sets of basic principles are discussed here.

The most fundamental set of principles in the formation, continuation, and change of all the diverse and amazing life forms on Earth, both past and present, are the processes and mechanisms of **genetic inheritance**. The nine principles discussed include:

- Principles of Genetics
- Principles of Opposites
- Principles of Randomness
- Principles of Competition
- Principles of Change
- Principles of Mathematics
- Principles of Molecules
- Principles of Reason
- Principles of Technology

An important thing in learning and appreciating the wonders of the heavens and life around us is the understanding that the belief God created the heavens and life on Earth does not require belief that creation was accomplished in six days by divine command.

Rather, belief that God created the heavens and all life on Earth calls for diligent efforts to learn and appropriately consider scientific evidence of the amazing ways and means used in creation and continuation of life.

It is logical to conclude in light of much scientific evidence that the Universe and life on Earth was created, is being sustained and being changed using many on-going identifiable natural principles over eons of time and not to believe that creation was all done by divine command in six days.

When a person sincerely believes that God created the Universe and all life on Earth, it is important to learn factually and honestly as much as possible about the **ways and means** by which creation was accomplished, the wonders of creation sustained and are continuing to change.

*

End of Essay

Background of Essay

The Evolution of Anti-Evolution in Public School Science Classes

— Glen E. Robinson

The young Charles Darwin served as official naturalist and geologist aboard the *HMS Beagle* during the British Navy's historic 1831-1836 around-the-world voyage for scientific discovery. A primary mission of the voyage was to map accurately the coast of South America. During the five-year voyage, Darwin copiously recorded observations and collected specimens of plants, animals, marine life, fossils and geologic formations. Darwin used these findings, along with later evidence, in development of his theory that living things had changed over extended time.

Working in South East Asia, half the world away, naturalist Alfred Russel Wallace independently developed his theory that living things had changed over time. In 1858, the Darwin and Wallace essays on the theory were published together, thus enabling both naturalists to share recognition for the theory, but neither essay received notable attention.

In 1859, Charles Darwin published his book, *On the Origin of Species,* which proposed the radically new theory that new species of plants and animals evolve over extended time due to small changes inherited from their ancestors and through a process he called "natural selection."

Most scientists of the day responded skeptically-positive and some responded enthusiastically to Darwin's theory, especially the biologists and zoologists. It provided scientists of the mid 19th century with scientific-based leads to some troubling scientific issues of the day. Scientists in many fields of scientific study began applying and validating Darwin's theory of Evolution, that life on Earth evolved over extended time by natural selection, to their field of scientific study, including: Botany, Zoology, Geology, Archaeology, Astronomy, Physiology, and Genetics.

Eventually, as more became known about the theory that life on Earth evolved over extended time by natural laws and principles of nature, the theory of Evolution became fundamental to virtually all scientific study. In the mid-20th century, genetic scientists discovered the actual genetic structure and mechanism of the evolutionary processes and began applying the findings in amazing and practical ways.

In 1952, James Watson and Francis Crick reported the startling discovery that genes are arranged in ribbons of genetic substances in different coded sequences wound together in the shape of the "double-helix." They were awarded the Nobel Prize for their tremendously significant findings. These discoveries have since become basic to all scientific study involved in genetic decoding and in its many applications including the monumental Human Genome Project started in 1990. With the assistance of super high-speed computers, the huge task was completed in 2003 with the mapping of all genes in the 3.1 billion DNA base pairs that make up the 23 pairs of human chromosomes.

In order to study genetic similarities among species, the genomes of several organisms were also decoded during the same period of time. Surprisingly, the genome of the mouse was found to be very similar to that of the human genome, a finding that has proved to be very helpful in the search for treatments of specific human diseases.

Thus, the validly of Darwin's basic theory that all life on Earth evolved over extended time by natural selection has been scientifically validated and appropriately applied countless times each day.

*

But early on, Darwin's book *On the Origin of Species* advocating the theory that living things evolved by small changes and natural selection over extended time provoked both anger and fear among persons who saw this as a dangerous attack on their fundamentalist theological belief that all creation occurred in one six-day week by Divine command as set forth in Genesis 1 in the *Holy Bible*.

In 1874, Charles Hodge, head of Princeton Theological Seminary, Princeton, NJ (not part of Princeton University) wrote a widely circulated book titled *What is Darwinism?* He answered that question bluntly: "It is **atheism** [and} utterly inconsistent with Scripture."

The issue was drawn—hard and narrow. The minds of true fundamentalist believers in the literal Divine six-day-creation should not be poisoned by hearing, reading, or considering anything said or written by "Godless Evolution Atheists" who believe in Evolution that teaches humans descended from monkeys and apes.

However, many persons believed that scientific investigations into how the Universe and life on Earth were created and maintained were urgently needed and should not be impeded by narrow-minded theological fundamentalists who insist that God created everything on Earth and in the Universe literally in just one six-day workweek. Therefore, those narrow-minded persons who denounce scientific findings and call scientists "Godless Evolution Atheists" should be ignored and scientists should go on with their important work of finding how the Earth and all life on it were formed.

*

The U.S. anti-evolution movement began in the 1920's. With the ending of WW1 in 1918, there was much public concern about political discrimination and the low state of social, religious and moral behavior. Several movements began to remedy these deplorable conditions through political action.

The 18th amendment to the US Constitution banning sale of alcohol was certified on January 16, 1919. The 19th amendment to the Constitution granting women the right to vote was ratified August 18, 1920. Tennessee's ratification was the final vote needed to add the 19 th amendment to the Constitution and passed the Tennessee legislature by a single vote.

Emboldened by successes in these hard-fought national reform campaigns, political activists turned their political know-how to state legislative action prohibiting the teaching of evolution in public schools and to passage of prohibitory laws varying among states and localities collectively called "Sunday Blue Laws." The Blue Laws banned public activities on Sunday such as: opening commercial businesses, showing movies, playing baseball games, etc.

Believing the teaching of evolution posed a major public danger, many religious fundamentalists sought state legislative remedy. The first anti-evolution law passed with little notice in Oklahoma in March 1923. Two months later, Florida adopted an anti-evolution resolution that had been proposed by William Jennings Bryan, three times U.S. Presidential candidate. In 1924, Bryan went to Tennessee and gave a speech in the state capital against teaching evolution; thousands of copies of his speech were distributed to state legislators and state residents. One year later, in March 1925, Tennessee became the third state to pass an anti-evolution law and the first state in which the law was tested in court.

The Tennessee test case was one of the most famous trials in American history. It took place in the small town of Dayton about 40 miles north of Chattanooga, in July 1925. On trial was a high-school teacher, John Scopes. The charge against him— **teaching Evolution.**

Note: The author wants the reader to know and to acknowledge with much gratitude to the Constitutional Rights Foundation that most of the information and descriptions concerning the historic Scopes Trial—the local settings, the trial participants, the legal procedures and the legal issues involved— have been borrowed directly or excerpted from the CRF authoritative publication: Bill of Rights in Action 22.2, titled: The Scopes Trial: Who Decides What Gets Taught in the Classroom?

The trial was staged by local business men to bring Dayton much needed publicity. It was widely promoted as the "Monkey Trial." News reporters came to the town in huge numbers. Townspeople organized a "Scopes Trial Entertainment Committee" to help arrange accommodations. The trial occurred in a carnival atmosphere. Hot-dogs and soft-drink stands lined the main street. Southern barbeque was kept roasting in a barbeque pit behind the courthouse. It was the first trial ever to be broadcast on public radio.

Two famous lawyers were to argue the classic "Inherit the Wind" case for and against the young local Science teacher accused of the crime of teaching evolution. Clarence Darrow, the famous trial defense lawyer, volunteered his services to the American Civil Liberties Union (ACLU) that was supporting Scope's defense, to serve as defense attorney without pay, the only time in his career that he did so. Darrow wanted to free people from the unthinking belief in biblical truth and encourage skepticism and scientific inquiry.

There was similar passion among the fundamentalists who believed that Darwinism undermined belief in the Bible. William Jennings Bryan, the prominent trial lawyer and nationally known politician, agreed to join the prosecution team. Bryan was a hugely popular orator who drew 4,000 people to his Sunday Bible classes at Royal Palm Park in Florida. Bryan went to Dayton not so much as a lawyer going to court, but as a preacher going to a revival meeting. He saw the Scopes trial as a "battle royal" in defense of the faith.

In preparing for trial, Darrow focused on discrediting the scientific validity of the anti-evolution law and encouraging scientific inquiry. To this end, he put together a group of eight distinguished scientists and theologians who would explain the scientific basis for evolution and show that it did not conflict with the Bible.

Bryan's strategy was far different. He would have liked to present scientific experts to show the flaws and gaps in evolutionary theory, but he could not find any distinguished scientists who would agree to testify. Instead, he focused on the argument for majority rule. In a letter to one of the prosecutors, he said, "This is the easiest case to explain I have ever found. The right of the people speaking through the legislature to control the schools which they create and support is the real issue as I see it."

Early in the trial, prosecution lawyers objected to the defense calling expert witnesses. They argued that expert testimony would be irrelevant because the law banned any teaching about human evolution. It did not matter whether or not it conflicted with the Bible or was scientifically valid. The judge agreed and ruled that the defense would not be allowed to present their expert witnesses to the jury.

The only issue that remained was whether Scopes had violated the law. But Darrow had one last strategy to show that the Bible could not be interpreted literally. On the last day of trial, he called Bryan as an expert witness on the Bible. Bryan, who had been teaching the Bible for years, could not resist.

The trial had been moved outside because the judge was worried that the courtroom floor might collapse. So Bryan took the stand on the courthouse lawn, surrounded by 2,000 people sitting on benches under maple trees and sitting on the grass.

What followed was a debacle for the witness. Darrow posed numerous questions about events recounted in the Book of Genesis. Did Jonah live inside a whale for three days? How could Joshua lengthen the day by making the Sun stand still? Bryan had no good answers to the questions, and interactions grew nasty. When lawyers tried to stop the questioning, Bryan shouted. "I am simply trying to protect the word of God against the greatest Atheist or agnostic in the United States."

"I object to your statement," Darrow shouted back. "I am examining your fool ideas that no intelligent Christian in the world believes."

After two hours, the judge adjourned the court. The next day, the defense conceded that it had no defense to the charge that Scopes had taught evolution.

The judge sent the case to the jury that in nine minutes returned a verdict of guilty. Scopes was fined $100, the maximum under the law.

ACLU appealed the case to the Tennessee Supreme Court on grounds the statute was unconstitutional. The court narrowly upheld the constitutionality of the statute, but overturned the verdict on a technicality, which ruled out any chance of taking the case to the U.S. Supreme Court.

Nether side achieved clear victory in the case. The jury found Scopes guilty, but his conviction was overturned. Bryan had taken a beating in court and widely ridiculed in the national press. Five days after the trial, he died in his sleep.

ACLU had brought the case to get a definitive ruling in favor of free speech and against the anti-evolution laws. The Scopes case failed to achieve this goal.

Two more states passed anti-evolution bills after the Scopes trial: Mississippi (1926) and Arkansas (1928). Various state and local school boards passed measures barring use of textbooks with materials on evolution. But the laws were never enforced, and the ACLU couldn't find anyone to challenge them.

Forty years later, Susan Epperson, a 9th grade biology teacher in Little Rock, Arkansas decided to take up the challenge. Epperson was teaching from a new edition of a textbook titled *Modern Biology*, which discussed the fossil evidence for human evolution and that "although man evolved along separate lines from primates, the two forms may have had a common generalized ancestor in the past." The text was in direct conflict with the Arkansas law that barred teaching "the theory or doctrine that mankind ascended or descended from a lower order of animals."

Backed by the Arkansas Education Association, she filed as complainant in December 1965, asserting that the anti-evolution law violated her freedom of speech and other constitutional rights. The case went to the U.S. Supreme Court in 1968. The Court based its decision on the First Amendment, specifically on its ban against the government establishing a religion and struck down the Arkansas anti-evolution law. This ended the first chapter in the legal debate over teaching evolution in public schools.

Chapter One could be appropriately titled: *Prohibiting the Teaching of Evolution*.

*

Chapter Two began and might be appropriately titled: *Requiring the Teaching of Creationism—Subtle Ways to Circumvent the Court Decisions.*

Advocates of teaching Creationism became ardently engaged in finding ways to get around the court decisions that Creationism is a **religious belief** and therefore cannot be taught as scientific based theory in public school Science classes. Two rounds of efforts have been tried so far and both found wanting. A third effort was soon in progress.

The first approach was to **rename** the Creation theology to be *Creation Science* and insist that it be taught in public school Science classes as an alternate scientific theory to that of the theory of Evolution.

In the early 1980s, several states attempted to introduce creationism along with the teaching of evolution, and the Louisiana legislature passed a law titled the *Balanced Treatment for Creation-Science and Evolution-Science Act.* The Act did not require teaching either creationism or evolution, but did require that if evolutionary science was taught then "creation science" must be taught as well. Creationists lobbied aggressively for the law. The stated purpose of the Act was to protect **"academic freedom."**

In connection with later "academic freedom" legislation, Louisiana Governor Jindal, who was a biology major during his time at Brown University, even received a strong veto plea from his former genetics professor, Arthur Landy saying, "**Without evolution, modern biology, including medicine and biotechnology, wouldn't make sense.**" Professor Landy later wrote, "I hope he [Jindal] doesn't do anything that would hold back the next generation of Louisiana's doctors."

In 1987, the U.S. Supreme Court found in the Edwards v. Aquillard case that *Creation Science* was not based on scientific evidence but was a **religious-based belief** that cannot be taught in public school Science classes.

*

The second approach was to **remove** any reference to religious theology from the issue and term the concept *Intelligent Design*. The approach was that the Universe is so vast, orderly and complex that it had to have been created through *Intelligent Design* and thus implying an anonymous *Intelligent Designer.*

On December 20, 2005, the U.S District Court for the Middle District in Pennsylvania ruled that *Intelligent Design* (ID) was not fact based theory of the origin of the Universe but a religious based belief similar to that of **Creation Science** and could not be taught as a science-based theory in public school Science classes. The Court also stated that this decision did not mean the ID should not continue to be studied, debated, and discussed in public schools. But most important, the court found that *Intelligent Design* **"cannot uncouple itself from its creationist, and thus religious, antecedents."**

The attorney for the plaintiff parents stated the court record would show the parents were entitled to more than $2 million but were going to accept less than half that amount in recognition of the small size of the school district and that the old school board had been voted out of office and the newly elected school board had to pay the huge bill. The new school board voted unanimously with one abstention, to not appeal the decision to the U.S. Supreme Court and to pay the plaintiff parents $1,000,000 in legal fees and damages. The school board accepted the decision of the federal district court and there was no appeal to the U.S. Supreme Court. Therefore, Kitzmiller v. Dover Area School District (2005) remains the final decision of the Federal courts and law of the Nation.

Judge John E. Jones III issued an extraordinary 139 page decision in the Kitzmiller v. Dover trial. In his Conclusion he wrote:

- The proper application of both the endorsement and Lemon tests to the facts of this case makes it abundantly clear that the Board's ID [*Intelligent Design*] Policy violates the Establishment Clause. In making this determination, we have addressed the seminal question of whether ID is science. We have concluded that it is not, and moreover that ID cannot uncouple itself from its creationist, and thus religious, antecedents.

· The citizens of the Dover area were poorly served by the members of the Board who voted for the ID Policy. It is ironic that several of these individuals, who so staunchly and proudly touted their religious convictions in public, would time and again lie to cover their tracks and disguise the real purpose behind the ID Policy. With that said, we do not question that many of the leading advocates of ID have bona fide and deeply held beliefs which drive their scholarly endeavors. Nor do we controvert that ID should continue to be studied, debated, and discussed. As stated, our conclusion today is that it is unconstitutional to teach ID as an alternative to evolution in a public school science classroom.

Judge Jones himself anticipated that his ruling would be criticized, saying in his decision that:

Those who disagree with our holding will likely mark it as the product of an activist judge. If so, they will have erred as this is manifestly not an activist Court. Rather, this case came to us as the result of the activism of an ill-informed faction on a school board, aided by a national public interest law firm eager to find a constitutional test case on ID, who in combination drove the Board to adopt an imprudent and ultimately uncon-stitutional policy. The breathtaking inanity of the Board's decision is evident when consid-ered against the factual backdrop which has now been fully revealed through this trial. The students, parents, and teachers of the Dover Area School District deserved better than to be dragged into this legal maelstrom, with its resulting utter waste of monetary and personal resources.

In fulfilling Judge Jones prediction, John G. West, Associate Director of the Center for Science and Culture at Discovery Institute, said:

The Dover decision is an attempt by an activist federal judge to stop the spread of a scientific idea and even to prevent criticism of Darwinian evolution through government-imposed censorship rather than open debate, and it won't work. He has conflated Discovery Institute's position with that of the Dover school board, and he totally misrepresents intelligent design and the motivations of the scientists who research it.

Newspapers of the day noted that the judge is "a Republican and a churchgoer." In the months following the decision, Judge Jones received death threats and he and his family were given around-the-clock federal protection.

The Kitzmiller v. Dover (2005) decision stated firmly that **Intelligent Design** was not a scientific based theory but religious theology that "**cannot uncouple itself from its creationist, and thus religious, antecedents.**"

This firm decision that teaching the religious theology of **Creationism** or any of its surrogates including **Intelligent Design** in public school Science classes violates one of the most cherished provisions of the U.S. Constitution—**The Separation of Church and State**—would seem to have ended the long saga of first, efforts to **prohibit** the teaching of the science-based theory of Evolution in public school science classes, followed by efforts to **require** the religious theology of **Creationism** and **Intelligent Design** to be **taught** as science-based in public school science classes to have closed and decisively ended the controversy. Wrong!

Instead of accepting the decision and moving forward with the development and support of sound, authentic science-based teaching for students in public schools during this **era** when quality science teaching and informed student interests in scientific study are so vital to America's national defense, global economic position, and domestic well being, some ardent promoters of **Intelligent Design** were enraged and denounced the decision. These proponents promptly moved to **confuse the legal issues**, to capture **control** of state-prescribed public school **science curriculums**, and to insert **Intelligent Design** into the science curriculum under the pretext of "**Academic Freedom**" and similar misleading guises.

Thus, **Chapter 3** opened with actions that can be justifiably titled "**Controversy, Deceit and Control**."

*

John G. West, Associate Director of the Center for Science and Culture at **Discovery Institute**, in his reply to the court's Dover decision, alluded to the deep anger and to the **Discovery Institute's** determined efforts **to defy** the court's decision, and by various means, to confuse the issues and to continue to force **Intelligent Design** into public school science classes.

At the time of his Dover decision, Judge John E. Jones was aware of what was termed the "**Wedge Strategy.**" This was set forth in a **manifesto** that described how, under the aegis of **Discovery**

Institute, Intelligent Design extremists intended to enlist other reactionary groups to join with them in a broad social, political, religious and academic **action-agenda** to defeat **materialism** and **naturalism**, with the defeat of **Evolution** to serve as the leading "Wedge" in the splitting action.

The Judge was obviously well aware of the Wedge Strategy that was already under way to undermine and subvert the impact of both the previous and the current Federal Court decisions regarding the teaching of the religious beliefs of Creationism and Intelligent Design in public school science classes. This apparently was one of the reasons for the lengthy court decision and the specific finding that Intelligent Design "**cannot uncouple itself from its creationist, and thus religious, antecedents.**"

This part of the ruling especially angered the **Creationism** and **Intelligent Design** promoters and caused them to attempt to reverse their public-image strategy from being viewed as **aggressors** attempting to force **Creationism and Intelligent Design** into the science curriculum of public schools, to that of being viewed as the **defenders** and **champions** of "**academic freedom**" for public school science teachers to be free to teach **Intelligent Design** as a **controversial** science issue in public school science classes.

*

In regard to the "**Wedge Strategy**" manifesto and related information about the **Discovery Institute** that were available to Judge John E. Jones when he issued the Kitzmiller v. Dover decision—for those persons interested in learning specifically about this information, online computer searches will yield a number of related documents. Just "Google" terms such as, *Wedge Strategy, Discovery Institute, Anti-Evolution Legislation, Timeline How Creationism has Evolved, etc.* and follow the leads.

For example, *Wikipedia, The free Encyclopedia* has a lengthy article on the history and operations of the "Wedge Strategy." The article states in part:

Note: This is verbatim copy with specific hyperlinks and endnote citations in the original. indicated in numeric order here.

The **wedge strategy** is a political and social action plan authored by the <u>Discovery Institute</u>, the hub of the <u>intelligent design movement</u>. The strategy was put forth in a Discovery Institute <u>manifesto</u> known as the **Wedge Document,**[1] which describes a broad social, political, and academic agenda whose ultimate goal is to defeat materialism, naturalism, evolution, and "reverse the stifling materialist world view and replace it with a science consonant with Christian and theistic convictions."[2] The strategy also aims to affirm God's reality.[3] Its goal is to change American culture by shaping public policy to reflect conservative Christian, namely <u>evangelical Protestant</u>, values.[4] The wedge metaphor is attributed to <u>Phillip E. Johnson</u> and depicts a metal <u>wedge</u> splitting a log to represent an aggressive public relations program to create an opening for the supernatural in the public's understanding of science.[5]

<u>Intelligent design</u> is the religious[6] belief that certain features of the universe and of living things are best explained by an intelligent cause, not a naturalistic process such as <u>natural selection</u>. Implicit in the intelligent design doctrine is a redefining of <u>science</u> and how it is conducted (see <u>Theistic science</u>). Wedge strategy proponents are opposed to <u>materialism</u>,[7][8][9] <u>naturalism</u>,[8][10] and <u>evolution</u>,[11][12][13][14] and have made the removal of each from how science is conducted and taught an explicit goal.[15][16] The strategy was originally brought to the public's attention when the Wedge Document was leaked on the Web. The Wedge strategy forms the governing basis of a wide range of <u>Discovery Institute intelligent design campaigns</u>.

Even with all the information online about the **Discovery Institute's Wedge Strategy** operations, it is difficult to think that any group of rational and well-intentioned Americans in this current scientific and global information age would have as their realistic goal to rollback a century of American scientific and cultural experience and to replace it with their own religious-based brand of pseudoscience "consonant" with their own "theistic convictions."

It is also surprising that the **current** Wedge Strategy promotion tactics are to revive and again misuse the old Scopes Monkey-Trial, fear-of-Evolution tactics used so unsuccessfully by William Jennings Bryan in Dayton, Tennessee nearly a century ago. This time, again the **villain** is **Charles Darwin** and the **fear** is **Darwinism.**

For example, the **Discovery Institute** website now annually sponsors February 12, the lampooned birthday of Charles Darwin, as "Academic Freedom Day." The current website version displays a ridiculous caricature of Darwin next to a cartoon drawing of the American Flag showing several arrows having been shot at the flag. This cartoon is accompanied by the following text:

> *On Charles Darwin's birthday (February 12th), students everywhere can speak out against censorship and stand up for free speech by defending the right to debate the evidence for and against Darwinian evolution. Let's make "Darwin Day" Academic Freedom Day!*

Across the country academic freedom on evolution is being trampled every day. Scientific research challenging Darwinism is thwarted. Teachers trying to discuss these challenges with their students are censored. Even students are subject to harassment and prejudice for expressing views that are skeptical of Darwinism. Scientists, educators, students need your help to protect their academic freedom rights. . . . Academic Freedom Day events can be as simple as having a table on campus where people can sign the Academic Freedom Petition and find out more about academic freedom on evolution. Or the events can be more elaborate including screening Expelled: No Intelligence Allowed or Icons of Evolution on campus.

Click here for more ideas on what you can do to celebrate Academic Freedom Day.

Academicfreedomday.com is hosted by Discovery Institute's Center for Science & Culture in partnership with the IDEA Center.

For more information about Academic Freedom Day and academic freedom issues e-mail us at academicfreedom@discovery.org.

The Intelligent Design extremists not only defy the Federal Court decisions that Intelligent Design is a theology-based **religious belief** and therefore cannot be taught in public school **science classes** but they also refuse to face the reality that modern, up-to-date genetic principles of Evolution are now the widely-accepted and central-unifying concepts for virtually all branches of legitimate science today.

It is incredible that any rational and well-intentioned group of people would promote their own brand of **pseudoscience** with the stated intention to confuse, to confound, to refute and to replace the basic and repeatedly validated findings and widely used procedures of the **authentic** sciences. Any attempt to rationalize such actions would be a ridiculous non sequitur.

It is important to remember that the Federal Courts have **already** examined the so called "scientific evidence" **claimed** by Intelligent Design advocates to support Intelligent Design and have not found **authentic** scientific evidence that the Universe and all life on Earth were created by supernatural command in a brief period of time. Thus, the Federal Courts more than once have **already** found Creationism and Intelligent Design to be **religious-based beliefs** that cannot be taught in public school Science classes in violation of the Constitution's First Amendment separating Church and State. It is unrealistic to believe that continued vociferous assertion that **pseudoscience** produces **scientific evidence** would cause the Court to overturn all the Court's previous carefully-considered and well-established scientific evidence to the contrary

It is deplorable that attempts would be made to deceive and dupe many individual American citizens, groups, businesses and institutions into supporting such a cause both **financially** and **politically** if those persons and groups realized that the activities they were supporting are actually part of a **conspiracy wedge** to breach the U.S. Constitutional wall separating Church and State.

*

The **Discovery Institute**—according to its company publications, company websites, press reports, and promotional and financial solicitation materials— is recruiting, training and financially supporting both full and part-time cadres of persons to be pseudoscientific specialists, media specialists, skilled attorneys, political strategists, and experienced legislative lobbyists—all ready to support or oppose public officials or candidates and to enact or defeat political issues or causes that are designated.

There are many implications for the health of American democracy of such a potentially large anti-scientific, ultra-extreme, highly-financed, and politically-powerful entity. These implications go far beyond the scope of this *Background Summary* of the long struggle to achieve and maintain sound, authentic, high quality science instruction for America's public school students. The broad social, political and religious implications must be left for other concerned citizens, groups and the authentic scientific communities to address.

*

The current tactics of the **Wedge Strategy** proponents to force Intelligent Design into public school science classes, in spite of the Federal Court decisions declaring Intelligent Design a nonscientific based religious-belief, include the following:

1. **Confuse** the issues!

2. Promote **controversy!**

3. **Change the legal issue** from Constitutional separation of **Church and State**, to **Academic Freedom** of teachers to teach **Intelligent Design** in public school science classes.

4. Reverse and change the public image from **aggressor** for Intelligent Design, to **defender** and **champion** of **Academic Freedom**, Critical Thinking and Freedom of Speech.

5. Declare anyone and all who dare to oppose the self-styled defenders of academic liberty to be **opponents** of

academic freedom, critical thinking, and freedom of speech.

6. Ignore all Federal Court decisions to the contrary. Focus on enacting **state-wide** legislation affecting control of public school **science curriculums** and **instructional programs** to permit, encourage, and protect the teaching of Intelligent Design or "Design Theory" under the guise of "**Academic Freedom**" within each specific state.

The **Discovery Institute's** news release regarding the recently passed laws in Tennessee and Louisiana illustrates the current maneuvers to do indirectly what the Federal Courts have found to be unconstitutional. Here are excerpts from the news release dated March 26, 2012:

> [Headline]—Tennessee Legislature Passes Landmark Academic Freedom on Evolution Bill.
>
> Bill Protects Teachers who Cover Controversial Science Subjects Objectively.
>
> Nashville-By a vote of 72-23, Tennessee's House of Representatives today passed an academic freedom bill that would protect teachers and school districts who wish to promote critical thinking and objective

discussion about controversial science issues such as biological evolution, climate change and human cloning.

"This bill promotes good science education by protecting the academic freedom of science teachers to fully and objectively discuss controversial scientific topics, like evolution," said Casey Luskin, science education expert and policy analyst at Discovery Institute's Center for Science & Culture. "Critics who claim the bill promotes religion instead of science either haven't read the bill or are putting up a smokescreen to divert attention from their goal to censor dissenting scientific views."

The bill expressly states that "it shall not be construed to promote any religious or non-religious doctrine."

The Tennessee State Senate previously passed the bill with overwhelming bi-partisan support. The Tennessee bill is similar to an academic freedom policy adopted in 2008 by Louisiana, known as the Louisiana Science Education Act.

This year, four states have considered academic freedom legislation designed to protect teachers who teach both scientific strengths and weaknesses of evolutionary theory. Many of the bills have been adapted from sample legislation developed by Discovery Institute, including a model statute posted online at www.academicfreedompetition.com.

At least nine states currently have state or local policies that protect, encourage, and sometimes even require teachers to discuss the scientific evidence for and against Darwinian evolution.

[Footnote] The work of Discovery Institute is made possible by the generosity of its members. Click here to donate.

Discovery Institute—Center for Science and Culture. 208 Columbia St.—Seattle, WA 98104.

*

It should be pointed out that the Discovery Institute's so called "Academic Freedom" crusade to "protect teachers" from reprisal for expressing their opinion or engaging in the study, debate, or discussion of Intelligent Design in public school classes is a contrived fallacy.

Judge John E. Jones III in the extraordinary 139 page, 2005 Kitzmiller v Dover decision explicitly wrote:

The citizens of the Dover area were poorly
served by the members of the Board who
voted for the ID [Intelligent Design] Policy.
...With that said, we do not question that
many of the leading advocates of ID [Intel-
ligent Design] have bona fide and deeply
held beliefs which drive their scholarly
endeavors. **Nor do we controvert that ID
should continue to be studied, debated,
and discussed. As stated, our conclusion today is that it is unconstitutional
to teach ID as an alternative to evolution
in a public school science classroom.**
[Emphasis added]

Thus, looked at in the larger context, Intelligent
Design and Creationism, like other religious beliefs
such as—Judaism, Christianity, Islam, Hinduism,
Buddhism,—can be "studied, debated, and discussed"
objectively and appropriately as academic topics
in suitable public school classes. However, since all
of these are religious-based beliefs and none are
scientific-based theories, to teach any of them as a
scientific-based alternative to the scientific-based
theory of Evolution in public-school Science classes
would violate the cherished First Amendment clause
of the Constitution that guarantees separation of
Church and State.

The implications of this Court distinction would seem about as clear, fair, and consistent with basic Constitutional principles as reasonable persons could expect and abide by.

*

It should be noted that there is evidence that the "Academic Freedom" crusade is currently the leading edge of the Wedge Strategy to muster both financial and popular support to circumvent the Federal Court decisions to the contrary and get Intelligent Design taught in public school science classes.

This persistent, confusing and shifting anti-scientific effort has been and is continuing to harm the scope and quality of science instruction in our nation's public schools. For example, the anti-scientific confusion in some states is harming the quality of science textbook content and limiting the number and potentially increasing the price of quality science textbooks available for adoption and use not only in those states but also in other states.

There is evidence that the confusing and shifting anti-scientific drag is taking a toll on the quality and quantity of persons preparing to teach science and to enter science-related occupations and reducing the interest and willingness of the public to support scientific related activities and science teaching.

The April 16, 2012 issue of *The Charlotte Observer* contained an infomercial by Duke Energy Company (one of the largest in U.S.) that stated:

> Studies show the U.S. is losing its stronghold as the world leader in science and technology. This is a result of an aging workforce combined with fewer young Americans choosing careers in science and technology. . . . Duke Energy is a proud sponsor of the North Carolina Science Festival. We understand the importance of supporting programs that expose students to a wide variety of science-related career paths. By fostering a growing interest for kids in the fields of science, technology, engineering and math, North Carolina can continue growing and producing skilled workers who bring new thinking and innovation to our lives.

*

Basic scientific illiteracy is inexcusable in a modern scientific global age. Lack of basic scientific knowledge and understandings negatively affects our nation's ability to defend ourselves and others militarily and to maintain our global scientific and economic leadership. Scientific illiteracy negatively affects our abilities to make well-informed public domestic decisions plus personal decisions including those related to health and product purchases.

Students in this Scientific Age have a right to know and public schools have the duty to teach in an unbiased, unadulterated, comprehensive, and professional manner the best scientific evidence available including scientifically sound evidence about **how** the Universe and all life on planet Earth came to be and **how** they are being sustained and continuing to change.

*

End of Essay Background

Notes

Notes

Notes

CPSIA information can be obtained at www.ICGtesting.com
Printed in the USA
BVOW06s1551230815

414613BV00010B/150/P